I0473178

THE ETERNAL LIFE

APPLIED MATH TO LIFE RESEARCH

RUI M.F. NASCIMENTO

ISBN-10: 1477604286
ISBN-13: 978-1477604281

TABLE OF CONTENTS

INTRO

Respecting above all, what we, several cultures worldwide, have in common, I think I have successfully explained in a mathematical way, as objectively as a scientific explanation can be, all those classical questions that Humanity has always pursued, such as:

•What are we?

•Where did we come from?

•Where are we going?

•How?

•Why?

Astrologists tell us about the present astral change to the Aquarium Era, a time where Humanity will be able to understand, love, share, generate opportunities to everyone.

According to them, we have been living in the Pisces Era, characterized by an ambiguity related to what concerns "Good" and "Evil". To turn things even worse, everyone has been considering themselves as the "good guy" and all the opponents as the "bad guys". Every fight is "in the name of God" and everyone begs Him help to kill their enemy "Once and for All". The bad news is all of their enemies do exactly the same thing. With this mentality many are those who feel the need to kill their brothers, believing they are killing "the Evil Once and for All", by following the allegory of sticking Him with a prop in his chest until he bleeds to death, or simply by letting Him agonize with torture procedures, inhuman suffering, nuclear bombs or chemical weapons, and so on.

It's now time, to present, discuss and approve a new mentality, new philosophical, political, legal, economical, social and moral values, all based on a new understanding of us as Human Beings in this World, in this life, as well as in all of those to come.

We have the privilege, the honor and the responsibility to generate the seeds of a new paradigm, to grant a Society of Knowledge to our sons and daughters, for the future generations to come, to our genes, to our Human Race. Prepare to leave behind all that you think you know because there are no absolute truths, the dogma is no answer and we shall be prepared for continued learning in life.

Are you ready?

From now on, your perspective of life will never be the same.

The Eternal Cycle of Life

Those who do not have the answers, nor have "the time" or the patience to look for them, state we only live once and it all ends with death, but of course, they cannot support any of these allegations. Some others support only Matter ends, but Spirit is immortal. That is closer, but not there yet. Humanity has learned "a lot" until the present time, the very beginning of this twenty first century, however has its knowledge dispersed and does not seem to have the capability to put the knowledge pieces together, making possible to understand the One Masterpiece of God, Allah, Dja, The Supreme Intelligence, The Great Architect of the Universe, (+∞, +∞), whatever you may like to call It.

This work I am presenting is possible by listening to everyone, obtaining therefore culture, and thinking, meditating, listening to nature's voice, as well as our interior voice, obtaining therefore wisdom.

Combining all together, culture and wisdom, we

manage to get the pieces of the puzzle together, and we can finally observe, understand and feel, completely astonished, the Universe laws and structure, us, and our job in it.

When you feel ready to understand "The Great Miracle of Life and Death", please turn the page. I must advise you though that you will never see the World the same way you did before.

Let's go.

THE ETERNAL CYCLE OF LIFE
(1 Corinthians 15:35-54; John 3:16 / 10:28)

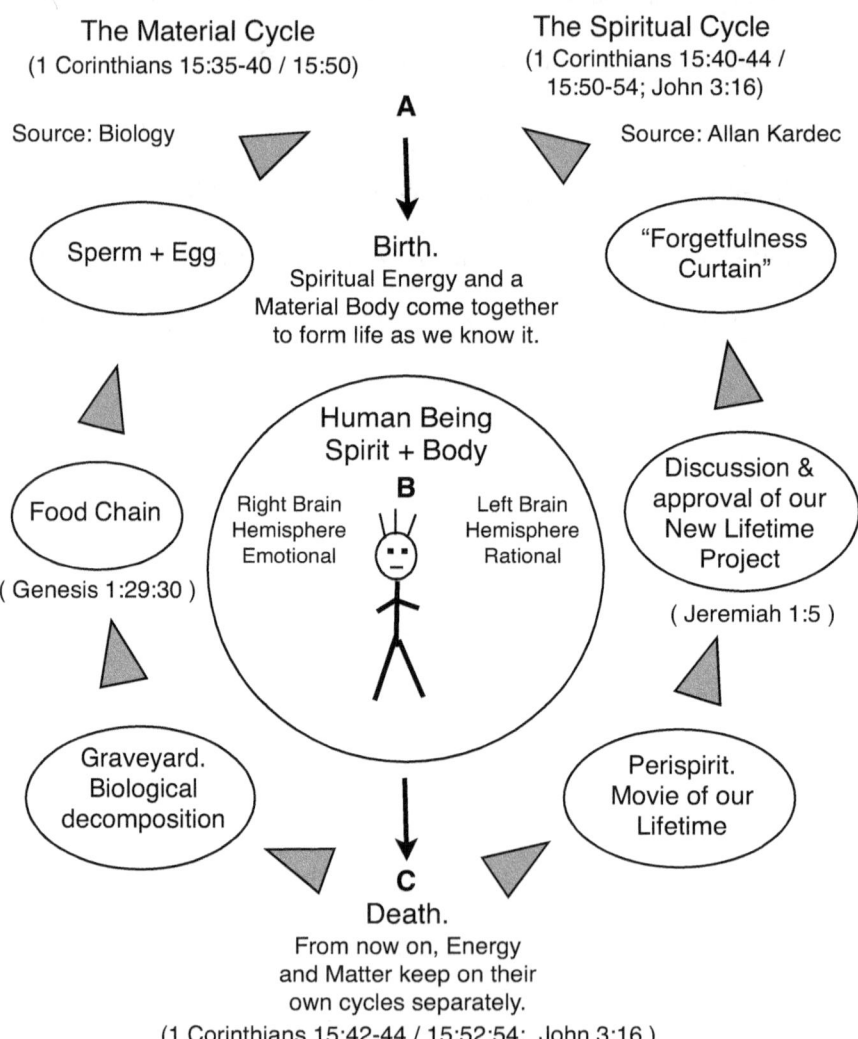

The Material Cycle
(1 Corinthians 15:35-40 / 15:50)

The Spiritual Cycle
(1 Corinthians 15:40-44 /
15:50-54; John 3:16)

Source: Biology

Source: Allan Kardec

A

Sperm + Egg

Birth.
Spiritual Energy and a
Material Body come together
to form life as we know it.

"Forgetfulness
Curtain"

Food Chain

(Genesis 1:29:30)

Human Being
Spirit + Body

Right Brain
Hemisphere
Emotional

B

Left Brain
Hemisphere
Rational

Discussion &
approval of our
New Lifetime
Project

(Jeremiah 1:5)

Graveyard.
Biological
decomposition

Perispirit.
Movie of our
Lifetime

C
Death.
From now on, Energy
and Matter keep on their
own cycles separately.
(1 Corinthians 15:42-44 / 15:52:54; John 3:16)

Human Being is Spiritual Energy on its Essence, having a material body as its outer shell.

It's Left Brain Hemisphere deals with science, logical thinking, numbers, can distinguish one thing from another.

It's Right Brain Hemisphere deals with emotions, feelings, premonitions, non-verbal communication & body language, telepathy and tends to see it as an all.

Most developed souls will be able to make a good use of both their brain hemispheres, on a balanced way.

Human Condition

Going back to the figure, we Human Beings get born on Stage A. We live our lives, on Stage B, according to what I will talk about on my next books and then one day we die, on Stage C. Our western civilizations cries the dead, some ancient civilizations use to have these big parties to celebrate the opportunity the soul just had to grow up. Most of us on our western civilization, do not have the slightest idea of what comes ahead next. Let me remind us all our Biology classes on the sixth grade to explain the Material Eternal Cycle.

Getting Born (Stage A)

The mail sperm and the female egg, when combined together, generate a fetus (Stage A), whom will generate a human body, more precisely, a baby.

Growing up, reproducing, evolving (Stage B)

This body will feed himself from another organic material, through his life (Stage B), by all those chemical reactions science is now able to explain. Food will allow the human being to generate new cells, replace the old ones, generate energy to be applied on his several activities.

Dying (Stage C)

Then, one day, we all die from the Human Condition (Stage C). Death means nothing more, nothing less, than the separation between our ethereal human essence we call spirit and our outer material shell we call material body.

The Material Eternal Cycle

The Graveyard (M1)

in most cases, our corpses get buried in a graveyard where some tiny animals known as decomposers will feed themselves from the dead bodies, decomposing it organically.

The Food Chain (M2)

These tiny animals are themselves food for bigger animals, and those for even bigger animals and so, the matter their bodies are made of, takes part on what science calls "The Food Chain" (M2). All the organic matter the decomposers leave on the soil, gets eaten by the vegetable life forms, witch will be eaten by animals or die and get decomposed again and again in the Food Chain.

Generating new life forms (M3)

By getting the necessary nutrients in the food chain, Humans and other sexual life forms produce sperm in males and eggs in females (Stage M3). From now on, you know the story: the male sperm and the female egg, when combined together, generate a fetus (Stage A), which will generate a human body, a baby.

We could go on and on, repeating over and over again
this story.

Why?

Because the Material Cycle is Infinite!

This one was easy, as most of us have learned it all in
school.

The Spiritual Eternal Cycle

As for the Spiritual Eternal Cycle, we were not that lucky as the following knowledge hasn't been taught in school, although is already known for many centuries ago. Why? Well, that will have to wait until my next books as that subject does not fit on this first book's theme. Anyway, although it has not been taught in school, it is one of those subjects you don't get via the regular means, but you can always search for it... the truth is out there... remember, curiosity is the main engine for increased knowledge. If you do not look for it, you cannot really complain that much, can you?

The Spiritual Cycle has not been taught at all to children in school nor to adults, but you will see than you can also put the pieces together here. Beginning with the part the majority of humans accept: the Human Being has its essence made of ethereal Energy, the Spirit, and has a material outer shell known as its body (Stage B). Trough life, the spirit climbs its Evolution Ladder by escalating one degree at a time on intelligence, courage, maturity, humbleness, altruism, generosity, joy, humor, etc.

Then, one day, the Human Being dies (Stage C), witch means it's now time for the spirit to leave the body. That dead body can no longer be useful to the progression of the spirit through the evolution ladder.

The movie of our lifetime (S1)

The spirit is now called a "Perispirit" by the Spiritual Doctrine and it's now time to head on to the next stage, where there will be an outer and self evaluation of this passed lifetime (Stage S1).

In those cases where the mind knows it's going to die, is prepared for it and accepts it, this journey is peaceful and everything goes smooth.

But in those cases where the human being was violently murdered and/or has some truth or crime to reveal before it goes away and/or has some feelings or messages to leave to someone, it knows it just cannot go away and leave it all like it is, it just has to do what it has to do, so it will try to contact with the "living" ones. "Living" is in quotes because material and spiritual cycles are eternal, so obviously, they can never die. "Living" is in quotes because that is the way we humans, refer to those who "live" now in the human condition, body and soul combined. So, these "tormented" spirits are now living in that other world, where they are a pure spirit, with no human body "attached" to them, and where only paranormal related professionals or curious can see them, hear them or reach them, as they have something really important to say. This S1 stage is the spiritual one all the horror movies talk about, as well as some religion's exorcisms, unfortunately with a whole lot of drama and no adequate information in quality or quantity. Until now, the common sense has been completely ignorant about this issue, and what Humanity cannot understand, it fears, it pursues, it condemns, it wants to "send away to Hell once and for all"!!!

That is precisely what some religions that didn't care for acquiring the necessary knowledge, and therefore have neglected it, have done with the practice of "exorcisms". Because those religions have neglected knowledge other doctrines studied seriously about this "outer" dimension, accusing them of being the Evil, those dogmatic religions made the worst thing they could do to those suffering spirits: they just did no care for them! They were never interested in knowing what it was all about, never cared for those spirits' cause of suffering, for their needs of revealing some unknown truth or personal message. These religions just called those spirits "Satanists", "The Devil", sending them away "Once and for All", no matter how long they tried to make themselves to be heard.

Confronted with the complete ignorance, superstition and unconsciousness of those priests, those "tormented" spirits try to spook every now and then someone they need to get attention from, trying to get themselves heard so their torment could finally cease. The look of a predator was always a defense mechanism due to the complete astonishment of those spirits towards all the misunderstanding, ignorance and lack of preparation of those "exorcists" to deal to such serious issues. Those cases where family or priests sent away the spirits were really only a delay, as the spirits left those places temporarily. You cannot force anyone to hear you, to love you or to care about you.

What do you do when you understand you are talking to such ignorant people who cannot seem to understand what you are talking about or doing? You just leave, you turn your back on them! But do you quit your goal?
Never!

You just wait for the chance to have someone who can listen to you, understand you, and help you to share the message you want to the people you need. And why do you need so desperately to send a message to someone in particular? Because in every single lifetime of yours, including this last one, all the suffering you lived and caused, all the love you lived and caused will be considered all the way from your birth, until the day you die. And you don't want to be judged by an error you made but have already regretted and are willing to smooth it the most you can, by ending some people's suffering telling them the truth. The truth of a situation, of a person, of an unsolved murder, of your love for someone who's unsure of it. You don't want to be judged by that because you still can solve the situation, or at least, part of it. Once the job is done, you finally accept your death in peace and head on to the next stage, where you will have a self and outer evaluation of this passed life of yours (Stage S2).

You and the angels who are judging you, have been watching "the movie" of this passed life of yours, where you are to recognize all the love and suffering you have lived and caused. "They", the Angels who are judging you, sure will lead you to recognize those actions because in that etherial world, you have no where to go, to run or to hide into, you are always accessible. Time is no longer like we knew it here on Earth and you've got all the time in the world to accept the facts, so you finally manage to understand the truth of your victories and defeats, of your pleasure and pain, of your love and hate, of your material assets and your spiritual characteristics.

Once all things get clarified, you can take your time before heading to the next stage. In some particular cases you can take decades or centuries, thousands of years too. It seems there is no average, every case is different, It all depends on your future Mission, on the time you need to get prepared and to get the necessary self confidence, as well as on the Earth's time you will need to wait in order to get the best context possible. Once we get there, it will be time to think about how our next life is going to be.

Discussion & Approval of your next Lifetime (S2)

We have already seen "the movie" of our last life, we have understood what we have done "good" and "bad", so we now have a realistic vision of our position in the graphic of life, that is to say, we know if we are in the first, second, third or fourth quadrant, we know if we are left or right wing, if we come from the centre towards an extreme side or if we come from an extreme position towards to the centre. The evolution of the spirit through the "Course Line of Life" is one and continuous, heading to $(+\infty, +\infty)$, we can never come back to live again previous degrees of evolution. We cannot either chose to live a degree of evolution several steps ahead of us, as that would be a perfect waste of time, we could not make it, it would affect negatively our self confidence, it would do more "harm" than "good", so that's just not possible in the Perfect Masterpiece of God, $(-\infty, -\infty)$, The Supreme Intelligence, The Great Architect of the Universe, Allah,

39

Dja, whatever. The ones who judge us and guide us through that stage have specific orders to do so, besides the fact that they don't control most of the system as well. It is a hierarchy, like we have in our corporations, family, social group, etc. This is the part of choosing our next life we can not choose by our own will. But there is a part we can choose.

Knowing we will have to live our next life in that specific degree of evolution, we can choose to live in the mountains or by the sea, we can choose to have this specific physical characteristic, but mainly, the most important of all, is that we do choose if we want to live a life extremely easy, easy, medium hard, hard or extremely hard, both from the material and spiritual points of view.

Choosing an easy life, you will not have to worry, things will happen naturally with no special drama, you will get to know the pleasures of material life, but you will take more years, which means, more lives to reach better worlds, meaning, to progress through the "Course Line of Life", towards $(+\infty, +\infty)$. Your progression will be slow and in the spiritual run to get to $(+\infty, +\infty)$, at the end of your lifetime, you will realize that you have wasted a lot of time, that you could have done a lot more progress, but you will remember with joy all the good and relaxing times spent with your friends and family.

Choosing a hard life, means you have wasted to much time in your last lives, and now, you have to catch up with those who were your mates, in a similar degree of evolution of yours, but meanwhile have attained advantage over you because they have worked harder, they weren't as lazy as you. You do not want to have a weaker performance than them, so you choose to live a hard life in your next lifetime. So you agree with the

Angels who are guiding you in that stage, that you will be born in a family and/or, in a society where people have a lot of difficulties to succeed in life. You agree that you will not have the same support of family and/or society, you will have to think, work, fight a lot more than the average of the family/society you will live in, you will have to "make the desert crossing" from that group you were born to the next group you want to live in. The group you were born lives in the degree of evolution you are presently in, and making the "desert crossing" will lead you to next step of evolution, the one you want to reach.

Once you have reached it, you will want to enjoy it, and you will, until it comes the day where you feel you have spent too much time enjoying and too little time progressing. Then you will choose a hard life, and so on and on, since $(-\infty, -\infty)$ until reaching $(+\infty, +\infty)$. That is the spiritual story of all of us. We all have to perform our

"Course Line of Life" from (-∞, -∞) until (+∞, +∞). Do you understand now, why the perfect God, (-∞, -∞), The Supreme Intelligence, The Great Architect of the Universe, Allah, Dja, whatever, gave to some of us an easy life, while he was giving another of us hard lives?

50% of our life's degree of difficulty was chosen by ourselves, the other 50% is defined by the degree of evolution our spirit lives in. How many times did you curse (-∞, -∞), The Supreme Intelligence, The Great Architect of the Universe, God, Allah, Dja, whatever, for your "bad luck" in this life, for the "injustice" of your life being more difficult than the ones around you?

(-∞, -∞), The Supreme Intelligence, The Great Architect of the Universe, God, Allah, Dja, whatever, have never failed, it was Humanity that could not reach the answers, simply because it was not the time yet, we didn't have the necessary education to understand it, because there were no global infrastructures that could be used for a global message spread, without having the censorship of the political, materialistic, dogmatic religious, conservative and spiritually unconscious elite! We all have to run the "Course Line of Life" one step at a time.

Several spiritual leaders were sent along the times to this world, to spread the news that the World Humanity knew, was about to end by the year 2000. Humanity understood that message as the end of the physical World we live in, the Planet Earth, because the physical dimension was the main one Humanity could understand. Spirituality was not really understood by the people, whom could only believe in what dogmatic religions prophesied or live in emptiness. These religions had an important role in Human evolution, as they had this difficult job of encouraging people to keep always their faith, although the first ones did never explain anything. We all have to realize that it was a truly hard and ungrateful job, but somebody had to do it. Dogmatic religions did it.

Meanwhile, a lot of errors were made, a lot of persecution, suffering and death took place, always "in the name of God, $(-\infty, -\infty)$, The Supreme Intelligence, The Great Architect of the Universe, Allah, Dja, whatever, but hey... That was nothing but "collateral damage". Spirits from third quadrant are in a primary state of evolution, do not have the conscience, intelligence, maturity, humbleness, etc., to understand things if someone came to explain them, and we have several examples of those messengers, such as Platan, Jesus Christ, Galileo Galilee, and many thousands more. The material body had to die anyway, so why not die in useful way to mankind and themselves? That's how we all have developed from the most primary stages of evolution until now.

The "Forgetfulness Curtain" (S3)

One of the main goals of being submitted to this forgetfulness curtain, is to forget a lot of information from that spiritual world, that is going to be an "outer" world when we reincarnate here, in this Human condition. That information is not deleted for good from us, it is kept in our subconscious, where accessing it does not depend on our conscious will, at least, the way we are used to until this early twenty first century. Another main goal of it, is to make us forget who our dear ones were, as well as our hated ones in our past lives. Can you imagine the chaos it would be if each one of us could revisit old dear ones, always trying to keep our ancient relationships, making impossible for them, as well as for ourselves, to live this lifetime, with our own objectives? We would all be stuck to the past, prisoners of our dearest memories, destroying all their newest relationships they would be trying to create in this life, necessary links to complete their present Mission? We would ruin this lifetime of our ancient dear ones and they would ruin ours. Would that be considered as love for them? It couldn't be. It would be an imperfection, an error of the system. And would it be good for ourselves? Of course not, we could not perform our job in this lifetime as well, and for our defense, we would say $(-\infty, -\infty)$, The Supreme Intelligence, The Great Architect of the Universe, God, Allah, Dja, whatever, made us like that, so it could not be an error of ours. And we would be right. Now imagine each one of us trying to seek revenge on those who harmed us in our last lifetimes. We would dedicate all our lives pursuing them just to make them pay for all that they had done to us. Realize that after watching the movie of our lifetime and understanding all that has happened to us, we would be aware of many things we had never got aware before, while "alive" in the human condition, so we definitely would have a lot to revenge. We all would leave our present lifetime project to later on, dedicating the moment to keep perpetual our dearest relationships, as well as pursuing and trying to destroy all of those who made us suffer or "killed" us in our last lifetime. I think you are getting the picture. Could it be worse? Sure, of course it could! If there was not a "Forgetfulness Curtain" in the system, that means it would not be only for the last life, but for all of them, lived in the past. Now, you try to figure the total chaos of each one of us, millions, billions of souls, trying not to miss, trying to cumulate our dearest relationships, the jealousy between our dearest ones disputing for our love, as well as our jealousy towards all the lovers the one we loved has loved through all of their lifetimes; every single one of us, trying to get revenged from all of those who caused us "harm"!!! it would be the total chaos!!! That's why (-∞, -∞), The Supreme Intelligence, The Great Architect of the Universe, God, Allah, Dja, whatever, cannot afford to neglect a "Forgetfulness Curtain".

After this "brainwash", each one of us is finally ready to incarnate in a "new lifetime", which in fact is the same spiritual life, the term means "new human identity" to be more precise, no more no less than it.

We have already made a deal with the Angels closer to (-∞, -∞) or (+ ∞, + ∞), The Supreme Intelligence, The Great Architect of the Universe, God, Allah, Dja, whatever, we already know what is going to be our job in our next lifetime. It's now time to take a deep breath, enjoy that spiritual world the most we can, take a last look around, exchange some impressions with all of those who live there, tell the last jokes... And get prepared to reincarnate in our next life... It will be painful... We all cry... They call it "Birth".

Mathematical Arguments

The Eternal Life Fallacy

This means this is not like they stand, Eternal Life does

not start after death, as that would mean

U=] Death, + ∞ [.

But then, having a defined starting point, it couldn't be

infinite, Eternal, could it?

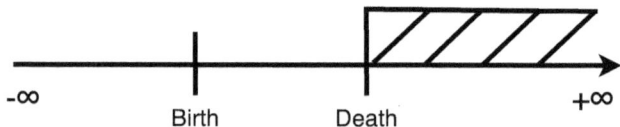

To be infinite, Eternal, it has to be like this: U =] -∞,+∞ [

(John 10:28)

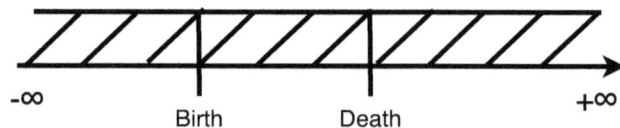

This means we were already living our Eternal Lives before we were born in this present lifetime, we are living it now, and we will proceed living it after our death from the Human Condition.

Allan Kardec supported on his work, "The Book of Spirits" that "The Spirit sleeps in a rock, dreams on a vegetable, excites on the animal and wakes up in the Human Being". And I ask what for?

Nobody wakes for nothing, just to be awaken. We all awake to work, to study, to relax, to do something, even if it's watching TV or to suffer in some alley gutter, to win, to loose, to be active or passive... there is no "nothing" in Universe.

Successive Lifetimes

My answer to my own question is "the Soul wakes up in Human Kind to be the next breed of living creature, the one that will leave planet Earth by the middle twenty second century, an alien, an extra terrestrial that will try to colonize maybe planet Mars, and with no Earth's gravity will suffer deep physical body changes in a couple of centuries.

Several Lifetimes

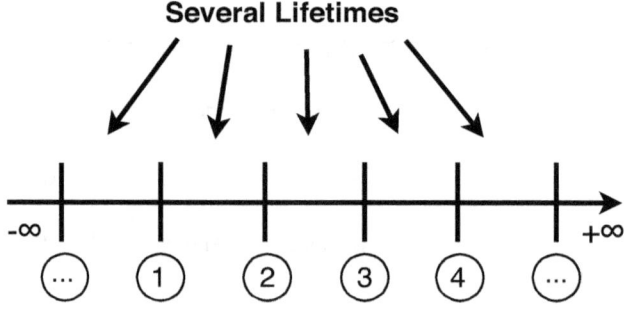

Each lifetime we get born precisely on the same point of spiritual, intellectual, emotional, material, etc. evolution we have died in our previous one.

So, there is no injustice in the World, we are in fact living a Perfect Masterpiece of Work of a Perfect Architect of the Universe, God, Allah, Dja, Odin, whatever, and each one of us is now the result of what we have developed in all our previous lifetimes. This may be hard to digest in the first place but is the deepest truth.

Longevity of Lifetimes through the Eternal Life

These spaces shown here as lifetimes aren't equal, both in quantity as in quality. Let's consider the spaces of several and consecutive lifetimes on a quantity basis:

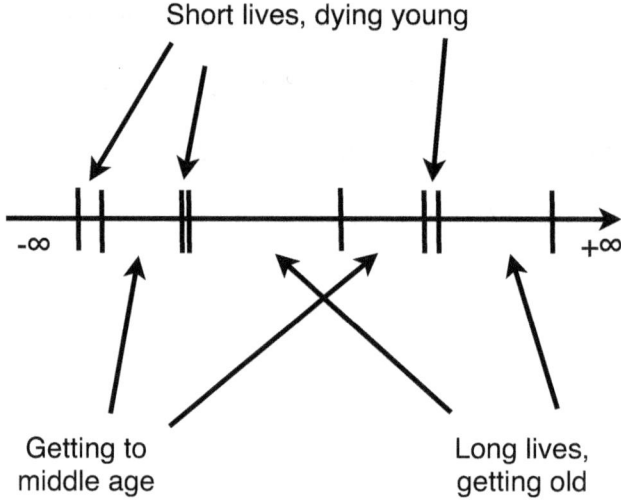

Along our Eternal Life, we live several lifetimes. In some of them we die young, in others we get to the middle age and in others we manage to get old.

The Age Fallacy

There has always been a misunderstanding regarding the age, the knowledge and experience of a single lifetime with the age, the knowledge and experience accumulated all along the Eternal Life.

For instance:

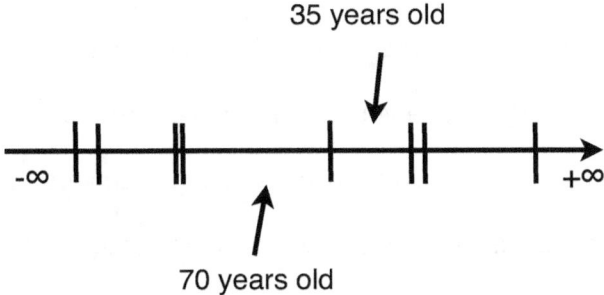

The old man says: "what do you know about life, young man? I am 70 years old, I have gathered a lot more knowledge and experience than you did... I've had your age a long time ago, I'm more than twice your age now! You still have to run a lot to even reach the point my feet stand on! You are too much irreverent, in this World you will never get old, you'll never get to my age, you'll die young!"

The young man replies: "yo, old man! Since when is the age of the material Body, directly proportional to the knowledge and experience of the intangible Spirit? Did you know, at the top of your 70 years old, that knowledge and experience are recorded in the inner spirit and not in the outer body? Have you ever learn that our brain is just an interface connecting the body nerve receivers to the intangible spirit? And for all about dying young, I just have to live my lifetime following my inner voice, my inner feeling. I'm not living someone else's life. What's the point of reaching 100 years old, if to do it, I'll have to live an empty life? Why should I? The scheme above shows you how we naturally die young in some lifetimes and naturally get old on others. So, why should I force living older one life I'm not supposed to? It would make as much sense as if another Soul would force dying young one lifetime it was supposed to get old... a complete non-sense!!! I want to live a full life, even if I'll die younger, I want and need to live the present lifetime as it is supposed to be... naturally. In the end, what I've learned and lived in shorter life must be greater and longer than what I could ever reach, sacrificing my goals just to live a longer lifetime.

Aren't we living an Eternal Life? So, what's the rush with the time issue? Eternal means there was no defined starting point nor will it be a defined ending point. By the way, having all our spirits living in $U=]-,\infty \ +\infty[$, which one of us is the oldest one?

Tell me something, old man: how come you are just counting the years, the knowledge and the experience gathered in one single lifetime, and not the years, the knowledge and the experience gathered all along the Eternal Life? How come you take into consideration only U={Present Lifetime} and not U=]-∞, +∞[?

That's your biggest mistake, old man!

Cheers... Love... Peace... Yo!!!"

Of course there are also souls living closer to (+∞, +∞) who manage to get old and souls dying young closer to (-∞, -∞), but this is the typical example of the misunderstanding young developed people had to face since ever, until the old men and women of their lifetimes manage to understand this content.

Regarding to the quality of each and consecutive lifetimes, I'll approach it in the next chapters.

And so, it's now closed our Material and Spiritual Cycles of the Eternal Life Chapter.

Now tell me... Is it so hard to understand the Eternal Life?

Bibliography

The Book of Spirits, Allan Kardec

Food Chain, Biology

The Holy Bible